Series 117

This is a Ladybird Expert book, one of a series of titles for an adult readership. Written by some of the leading lights and outstanding communicators in their fields and published by one of the most trusted and well-loved names in books, the Ladybird Expert series provides clear, accessible and authoritative introductions, informed by expert opinion, to key subjects drawn from science, history and culture.

The Publisher would like to thank the following for the illustrative references for this book:
Pages 9, 11: © Bettmann/Gettyimages; page 29: courtesy of CERN.

Every effort has been made to ensure images are correctly attributed; however, if any omission or error has been made please notify the Publisher for correction in future editions.

MICHAEL JOSEPH

UK | USA | Canada | Ireland | Australia
India | New Zealand | South Africa

Michael Joseph is part of the Penguin Random House group of companies whose addresses can be found at global.penguinrandomhouse.com

Penguin
Random House
UK

First published 2018
001

Printed in Italy by L.E.G.O. S.p.A.

A CIP catalogue record for this book is available from the British Library

ISBN: 978–0–718–18784–2

www.greenpenguin.co.uk

Penguin Random House is committed to a sustainable future for our business, our readers and our planet. This book is made from Forest Stewardship Council® certified paper.

Big Bang

Marcus Chown

with illustrations by
Chris Moore

Ladybird Books Ltd, London

The day without a yesterday

The greatest discovery in the history of science is that there was a day without a yesterday. The universe has not existed for ever. It was *born*. Around 13.82 billion years ago all matter, energy, space – and even time – erupted into being in a titanic fireball called the Big Bang. The fireball began expanding and, out of the cooling debris, there eventually congealed the galaxies – great islands of stars of which our Milky Way is one among an estimated 2 trillion others. This, in a nutshell, is the Big Bang theory.

Whatever way you look at it, the idea of the universe popping into existence out of absolutely nothing is utterly bonkers. For the average person it immediately prompts questions such as: What was the Big Bang? What drove the Big Bang? And what happened *before* the Big Bang? The last is the most awkward question of all. It was largely because of the need to answer it that most scientists had to be dragged kicking and screaming to the idea of the Big Bang. What forced them to accept it was one thing, and one thing only: evidence. And the evidence, it turns out, is literally all around us . . .

The heat of the Big Bang

The Big Bang fireball was like the fireball of a nuclear explosion. The heat of such an explosion, however, eventually dissipates into the surroundings. But the universe has no surroundings. By definition, the universe is all there is. So the heat of the Big Bang fireball had nowhere else to go. It was bottled up in the universe. Consequently, the heat 'afterglow' of the Big Bang is still around today. Greatly cooled by the expansion of the universe, it appears not as visible light but principally as a type of invisible light known as 'microwaves'. TV aerials pick up microwaves. If you were to tune an old 'analogue' TV between the stations, 1 per cent of the 'static' on the screen would be from the Big Bang.

The afterglow of the Big Bang fireball accounts for 99.9 per cent of all the particles of light, or 'photons', in the universe, with the light from the stars and galaxies accounting for only 0.1 per cent. If you had eyes that could see microwaves, the whole universe would appear to be glowing brilliant white. It would be like being inside a giant light bulb. But, although the afterglow of the Big Bang is the single most striking feature of the universe, it took a long time to realize we live in a Big Bang universe.

The Big Bang – Einstein's blind spot

Before 1915 all ideas about the origin of the universe had no more factual basis than fairy stories. After 1915 it was possible to come up with a scientific explanation. What changed everything, in November of that year, was the announcement by Albert Einstein of his theory of gravity – the general theory of relativity.

Einstein's genius was to realize that, although we are unaware of it, a mass like Earth creates a valley in the space-time around it. Other masses – like you and me – try to fall to the bottom of the valley, but the Earth's surface gets in the way. To explain our tendency to fall we have invented a 'force' called gravity that pulls us 'down'. But there is no such force. We are simply responding to warped space-time.

In 1916 Einstein applied his theory to the biggest mass he could imagine – the universe – and created 'cosmology', the science of the origin, evolution and fate of the universe. Because he was wedded to Isaac Newton's idea that the universe had no beginning and looks the same for ever, he proposed that empty space exerts a mysterious repulsive force that nullifies the attractive force between masses and keeps the universe 'static'. Thus he missed the message in his own theory.

But others did not.

The restless universe

To simplify his complex theory so that he could deduce something about the universe, Einstein assumed it had looked the same at all places and at all times. In 1917, however, the Dutchman Willem de Sitter dropped this assumption, proposing that the 'density' of the universe changed with time. From this he concluded that space must be either shrinking or expanding. The only trouble was the universe he imagined contained no matter and so was not realistic.

In 1922 the Russian Aleksandr Friedmann found that the universe would still be shrinking or expanding even if it contained matter (which, of course, it does). And this discovery was made independently in 1927 by the Belgian Catholic priest Georges Abbé Lemaître. Today Friedmann–Lemaître universes are more commonly known as Big Bang universes.

Einstein was wrong. The universe is not static and unchanging. We live in a restless universe. In an expanding Big Bang universe every mass flees from every other mass. And the further apart two masses are, the faster they flee, with masses that are twice as far apart as others receding at twice the speed, three times as far apart, three times the speed, and so on.

However, Big Bang universes were only theoretical until observations caught up.

The expanding universe

In 1923 the American Edwin Hubble turned the biggest telescope in the world – the 100-inch on Mount Wilson in Southern California – towards the Great Nebula in Andromeda. This was the largest and closest spiral-shaped smudge in the night sky. He wanted to know whether such 'spiral nebulae' were clouds of shimmering gas within the Milky Way – the great island of stars to which the Sun belongs – or separate stellar islands beyond the Milky Way whose stars were blurred together by distance.

The key was seeing individual stars. In 1929, Hubble spotted a special class of stars whose speed of brightening and fading indicated their distance. Hubble concluded that Andromeda was far beyond the Milky Way. He had discovered the building blocks of the universe: 'galaxies' – great islands of stars of which we now know our Milky Way is one among trillions.

Next, Hubble deduced how fast his galaxies were moving. The frequency, or pitch, of their starlight changes just as a police siren becomes shriller as it approaches and deeper as it recedes. From this 'Doppler shift', Hubble's fellow astronomer Vesto Slipher had found that some galaxies were fleeing from us at immense speed. Extending his work, Hubble found that almost all galaxies were receding: the greater their distance from us, the faster they were fleeing. We live in an expanding universe.

However, it took a Ukrainian-American physicist to realize the implications for the universe's origin.

The hot Big Bang

In the 1940s, George Gamow was looking for a place to forge the elements. There are ninety-two naturally occurring ones, from the lightest, hydrogen, to the heaviest, uranium. Once, it was assumed they had all been put in the universe on Day One by a Creator. But there was scientific evidence the universe had begun with only one element – hydrogen – and that heavier elements had been built up step by step from this basic building block.

The cores, or 'nuclei', of atoms repel each other fiercely, so must be slammed together violently to combine. This requires extreme heat, temperature being a measure of microscopic motion. The obvious furnace was inside stars, but stars were not thought to be hot enough. (This turned out to be a mistaken idea.) If not stars, then where?

Gamow pictured the universe's expansion running backwards like a movie in reverse. Eventually, everything was squeezed into a tiny volume: the Big Bang. But matter, when squeezed, gets hot, as anyone who has squeezed air in a bicycle pump knows. Gamow concluded that the Big Bang was therefore a *hot* Big Bang.

He calculated that such a furnace would turn 10 per cent of the universe's hydrogen into helium, the second lightest element. Unfortunately, it would also cool quickly and cease element-building before forging any elements much heavier. Although the source of most of nature's elements remained a mystery, Gamow had nevertheless stumbled on the idea that the Big Bang had been hot. And a hot Big Bang had consequences . . .

Afterglow of creation

George Gamow realized that, if the Big Bang had been a blisteringly hot fireball, then its relic heat must still be around in today's universe, greatly cooled by the expansion so that it appeared as microwaves – a form of radio waves. However, he thought that this 'afterglow of creation' would be indistinguishable from microwaves coming from other astronomical sources such as stars. He was wrong.

Gamow's students Ralph Alpher and Robert Herman realized that the afterglow of the Big Bang would have two features that would make it unmistakeable. First, it would be coming equally from every direction in the sky. Second, and more technically, its brightness would vary with energy like a glowing body paradoxically known as a 'black body'. Specifically, they calculated that the afterglow would correspond to a body that had a temperature of just 5 degrees above absolute zero, the lowest temperature possible, or −268°C.

Alpher and Herman published their prediction in the international science journal *Nature* in 1948. They even went to radio astronomers to ask whether the Big Bang microwaves might be detectable by radio telescopes. They were told (incorrectly) that they would not be. And so Alpher and Herman's prediction was forgotten for seventeen years.

The microwave glow of pigeon droppings

Arno Penzias and Robert Wilson were scientists employed by Bell Labs, a research facility in America run by the AT&T phone company. They had been lured to Bell in 1964 by the prospect of using a microwave 'horn' – a funnel the size of a train carriage – for astronomy. The horn, at Holmdel in New Jersey, was being used by AT&T's engineers to transmit and receive signals from Telstar, the first modern communications satellite. Once the tests were done, the horn was handed over.

Penzias and Wilson planned to pick up the ultra-faint microwave hiss of gas around our Milky Way. But everything emits microwaves – people, buildings, trees, even a microwave horn. So first they measured the spurious contributions, so as not to mistake them for what they were looking for. Unexpectedly, they were left with a persistent hiss of static, typical of a body at 3 degrees above absolute zero.

They thought the hiss was from New York, but they got it *wherever* they pointed the horn. They thought it was from within the Solar System, but it was unchanged as the Earth circled the Sun. Finally, they noticed two pigeons nesting in the horn. They had covered its interior with droppings, which would be glowing with microwaves. They ejected the pigeons and cleaned out the droppings. But the hiss remained. In the spring of 1965, at his wits' end, Penzias phoned a fellow scientist . . .

'Boys, we've been scooped!'

Penzias's phone call actually concerned another matter but, inevitably, he complained about not being able to do any astronomy with the Holmdel horn because of the annoying hiss. His friend mentioned a recent talk he had attended given by Jim Peebles of Princeton University.

Peebles' boss, Bob Dicke, believed in an oscillating universe, one that repeatedly expanded from Big Bangs and contracted down to 'Big Crunches' like a giant beating heart. By now it was known that stars *were* hot enough to forge the elements, so each cosmic cycle could begin the same way only if the elements made in the previous cycle were destroyed. Heat, Dicke realized, would do the job. Thus, like Gamow, he came to the hot Big Bang, though for the opposite reason – because he wanted to find a way to destroy elements rather than create them. According to Peebles, the Princeton team were planning to use a microwave horn to detect the heat afterglow, and were in the process of building one.

Penzias phoned Dicke. When he heard Penzias's story, he turned to his team, who were eating lunch in his office, and said: 'Boys, we've been scooped.' Incredibly, a microwave horn just thirty miles from Princeton had accidentally found what Dicke's team was looking for. For discovering the 'cosmic background radiation' and confirming the Big Bang, Penzias and Wilson would share the 1978 Nobel Prize in Physics.

A knock-out blow for the steady state

Despite the fact that the universe was expanding and so must have been hotter in its past, and despite relic heat being found, the idea that the universe was born in a Big Bang was not accepted immediately. For two years after their discovery, Penzias and Wilson never mentioned a cosmic origin for their anomalous hiss. Like most scientists, they believed instead in the steady state theory.

This was proposed in 1948 by British cosmologists Fred Hoyle, Hermann Bondi and Thomas Gold. It postulated that, as the universe expands and galaxies fly apart, matter fountains into existence in the gaps, condensing to form new galaxies, so the universe always looks the same overall. Matter, rather than appearing in a one-off Big Bang, is 'continuously created'. Such a universe was appealing because it had existed for ever, avoiding the awkward question of its origin.

The steady state predicted that the universe in the past looked the same as today. The light from distant objects reveals things taking place in the past since it takes billions of years to reach us. But observations showed that, just as the ancient Earth contained dinosaurs that no longer exist, the early universe contained super-bright galaxies called 'quasars' that no longer exist. Such evidence of an 'evolving universe', taken with the cosmic background radiation, had by the late 1960s killed the steady state, forcing acceptance of the Big Bang.

Quasers, powered by matter sinking into giant black holes, existed in the past, but not today.

Why wasn't the Big Bang afterglow found earlier?

Why was the afterglow of the Big Bang found only in 1965, and then by accident, when it had been predicted seventeen years earlier? One reason is that the prediction came from Gamow's discredited theory that all elements were forged in a hot Big Bang. Another is that, before 1965, it was extraordinarily difficult for scientists to take seriously any theory of the early universe. The temperature and density of matter was so extreme and so far removed from the everyday as to be unimaginable. According to Nobel Prize winner Steven Weinberg: 'The mistake of physicists is not to take theories too seriously, but not to take them seriously enough.'

But not only was the fireball radiation predicted before it was detected, it was also detected before it was detected! In 1938 American astronomer Walter Adams spotted the characteristic signature of molecules known as cyanogen floating in the space between the stars. Such molecules act like tiny thermometers, quietly taking the temperature of space. The problem was, they were not recording absolute zero, as would be expected in deep space, but 2.3 degrees above absolute zero. According to Canadian astronomer Andrew McKellar, something mysterious was heating them up. Only in 1965 did it become clear what the mysterious something was: the afterglow of the Big Bang.

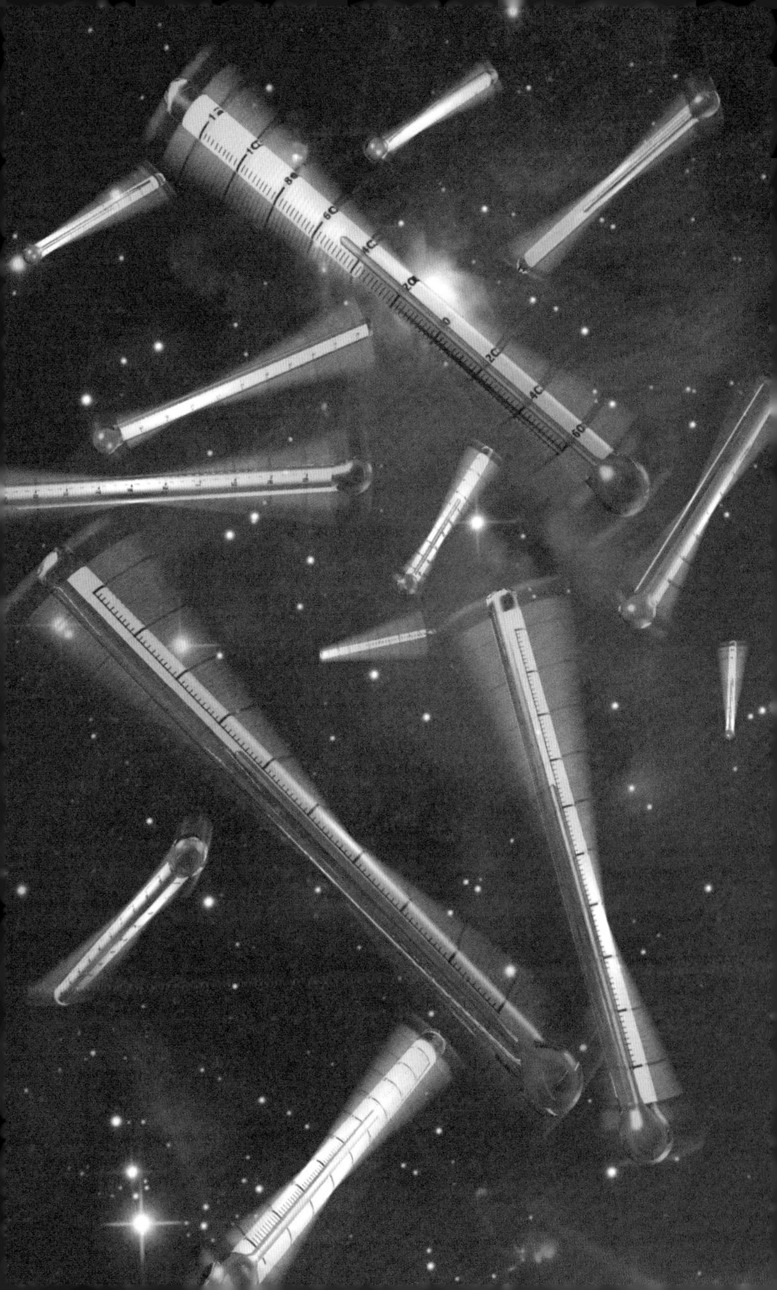

Where did the Big Bang happen?

If the universe exploded into existence in a fireball 13.82 billion years ago, an obvious question is: *Where* did the Big Bang happen? Here we are undone by the term Big Bang itself, which, ironically, was coined in a BBC radio interview in 1949 by Fred Hoyle, who advocated the steady state universe and never believed in the Big Bang.

The fireball picture painted by the term Big Bang is wrong in almost every respect. In a conventional explosion of, say, a stick of dynamite, the detonation occurs in one location and the shrapnel flies outwards into void. In the Big Bang there was no centre and no pre-existing void. Space itself popped into existence and began expanding everywhere at once.

Astronomy books often liken the universe to a rising cake, with raisins representing galaxies. As the cake grows, every raisin recedes from every other one. There is no centre of the expansion. Unfortunately, a cake has an edge, unlike the universe, which may go on for ever. All such analogies therefore fail to convey the true nature of the universe. They allow us only partial glimpses. As a four-dimensional object, the universe is fundamentally unimaginable by mere three-dimensional human beings. It is properly described only by the mathematics of Einstein's theory of gravity.

The Big Bang connects the big to the small

If, once again, we imagine the universe's expansion running backwards like a movie in reverse, it becomes ever denser and hotter. To understand what was going on closer and closer to the moment of creation when everything popped into existence, it is therefore necessary to understand the physics at ultra-high temperatures, which is synonymous with ultra-high energies. This is why machines like the Large Hadron Collider near Geneva in Switzerland, which accelerate subatomic particles such as protons to ultra-high energy then slam them together, are said to recreate the fiery conditions of the Big Bang at the collision point for a split second.

There is a limit to the high energies that can be probed at the LHC. However, out of the cooling fireball of the Big Bang, there eventually congealed the galaxies we see today. They are not spread evenly through space but form great daisy chains and sheets, wrapped like Christmas lights around enormous dark voids. And this pattern is believed to be a fossil imprint of processes going on in the Big Bang.

The universe can therefore tell us about microscopic physics just as microscopic physics can tell us about the universe. And this connection exists for the simple reason that, in a Big Bang universe, what is now big was once small.

Looking up at the 15-metre tall, 14,000-tonne Compact Muon Solenoid, a detector at the Large Hadron Collider, near Geneva, used for studying the fundamental building blocks of matter.

What's the matter with antimatter?

In the earliest moments of the Big Bang, when the temperature was enormously high, we have no idea what exotic subatomic particles were around. But, after a while, it became cool enough that the fireball consisted of familiar particles like protons and electrons, the constituents of present-day atoms from which all matter is made. Only there is a puzzle. It concerns antimatter.

Particles and antiparticles have opposite properties, such as electric charge. For instance, the antiparticle of the negative electron is the positive positron. Every physics processes we know of creates equal amounts of matter and antimatter. So why is the universe made of matter, and not equal amounts of matter and antimatter?

A clue comes from the fact that the afterglow of the Big Bang contains 10 billion photons for every particle of matter in the universe. Now, when a particle meets its antiparticle, it 'annihilates', ultimately into photons. The 10-billion-to-1 ratio can therefore be explained if the early universe contained 10 billion plus 1 particles of matter for every 10 billion of antimatter. After an orgy of annihilation, one particle of matter was left for every 10 billion photons.

But why did the laws of physics create a Big Bang with an ever-so-slight excess of matter over antimatter? That's one of the biggest puzzles in science.

Cooking up the elements in the Big Bang

When the universe was about three minutes old and its temperature had dropped to about a billion degrees, it was cool enough to build elements. Along with a vast number of photons, the Big Bang fireball contained a smattering of protons, neutrons and electrons – and the protons and neutrons started sticking to make the cores, or 'nuclei', of atoms.

It was a race against time. Free neutrons disintegrate, halving their number every ten minutes. And the fireball was expanding and cooling so fast that protons and neutrons were soon colliding so rarely and softly that they no longer stuck. By the time the universe was just twenty minutes old the orgy of element-building ended. Calculations show that 90 per cent of matter was left as the lightest element, hydrogen, and 10 per cent as the second lightest, helium. This is almost exactly what we observe today and is proof they were indeed forged in the cooling Big Bang fireball.

George Gamow claimed: 'The elements were cooked in less time than it takes to cook a dish of duck and roast potatoes.' But all heavier elements have been made since the Big Bang. The iron in your blood, the calcium in your bones, the oxygen that fills your lungs each time you take a breath, were all forged inside stars.

P
N
N
P
P
N
P
P

The epoch of last scattering

When the universe was 380,000 years old, and its temperature had dropped to about 3,000°C, it was cool enough for electrons and nuclei to combine to form the first atoms. This was a key moment in the history of the universe for several reasons.

Free electrons interact readily with photons, particles of light, so before the universe was 380,000 years old the 10 billion photons which were around for every particle of matter blasted apart matter before it could form clumps. Once electrons were safely locked away inside atoms, however, they tended to ignore photons. This meant that, for the first time in history, gravity could pull together matter and create the 'seeds' of cosmic structures such as clusters of galaxies.

When electrons had been free, they constantly 'scattered' photons, causing them to zig-zag crazily across space. Water droplets do a similar thing to photons in a fog, so the universe was opaque. Once electrons were locked away in atoms, however, the universe became transparent. For the first time, photons could travel in straight lines. And the photons of this 'epoch of last scattering' have been travelling unhindered across space for 13.82 billion years before they reach us as the cosmic background radiation.

Looking back in time with our telescopes, we see the universe emerging from the fog of the Big Bang like a tanker emerging from the fog at sea.

NORTH STAR
MONROVIA

A baby photo of the universe

Recall that, if your eyes could see microwaves, you would see the whole sky glowing white with the afterglow of the Big Bang, like the inside of a light bulb. (Strictly speaking, you would have to go into space to see this since the atmosphere glows with microwaves.) Well, actually, the glow varies ever so slightly because, when matter at the epoch of last scattering began to clump for the first time, it left its imprint on the sky.

In 1992 NASA's Cosmic Background Explorer (COBE) satellite found that the temperature of the afterglow of the Big Bang varied across the sky by a few parts in 100,000. An image of the sky was created. It was over-hyped. Stephen Hawking called it the 'greatest discovery of the century – if not of all time'. COBE scientist George Smoot said: 'It's like seeing the face of God.' What COBE had actually done was snap a baby photo of the universe, aged just 380,000 years.

COBE's hot and cold spots were the seeds of structures bigger than anything we have yet seen in today's universe. But in 2001 and 2009, respectively, NASA's Wilkinson Microwave Anisotropy Probe (WMAP) and the European Space Agency's Planck probe obtained much sharper baby photos. In effect, they saw the seeds of the galaxy clusters we see all around us.

The cosmic Dark Age

After matter began congealing at the epoch of last scattering, the Big Bang fireball continued to cool and fade until its light was no longer visible to the human eye (not that there was anyone around to see it). The universe, which had begun in a ferocious fireball, was plunged into darkness. And this cosmic 'Dark Age' was a prolonged one, stretching for hundreds of millions of years.

The universe grew and grew. It doubled in size. It became ten times as big. It expanded by a factor of 100. All in total darkness. And then, when the universe was about 400 million years old, something extraordinary happened. All across the universe, like some cosmic firework display, lights switched on. The first stars were born as gravity squeezed and heated clumps of matter sufficiently to ignite nuclear reactions. At about the same time there appeared 'quasars', giant black holes in the hearts of newborn galaxies which heated matter to such extraordinary temperatures as they sucked it in that they shone as bright as a hundred modern-day galaxies.

This dramatic end of the cosmic Dark Age has never been seen, only inferred from its imprint on the afterglow of the Big Bang. But everything could change when NASA launches into Earth's orbit its James Webb Space Telescope, capable of observing the greatest firework display in history.

The first Big Bang bolt-on: dark matter

The basic Big Bang idea – that the universe began in a hot dense phase and has been expanding and cooling ever since – is pretty incontrovertible. It is supported by the fact the universe is expanding, is permeated by left-over heat and has 10 per cent of its atoms in the form of helium. However, the basic Big Bang model contradicts observations in several major ways, which have meant bolting on new things.

Most seriously, the Big Bang predicts we should not exist! To understand why, it is necessary to know about galaxy formation. In the Big Bang fireball, regions which were slightly denser than others had slightly stronger gravity and dragged in more matter than others. In a process akin to the rich getting ever richer, galaxies grew year on year. Unfortunately, it would take much longer than 13.82 billion to make a galaxy like the Milky Way.

Astronomers fix this by postulating that the visible stars and galaxies are outweighed by a factor of six by invisible 'dark matter', whose extra gravity speeded up galaxy formation and so enabled a galaxy like the Milky Way to form. The composition of dark matter is a mystery. It could be made of novel subatomic particles or mini black holes. As yet, however, dark matter has not turned up in any terrestrial experiment.

The second Big Bang bolt-on: dark energy

Another way the basic Big Bang contradicts observations is in predicting something incorrect about cosmic expansion. As Hubble discovered, the galaxies are flying apart like pieces of cosmic shrapnel in the aftermath of the Big Bang. The gravitational attraction between galaxies acts like a web of invisible elastic and should be slowing down cosmic expansion.

However, in 1998, astronomers discovered that the universe's expansion is actually speeding up. They fixed the problem by postulating the existence of 'dark energy'. This is invisible, fills all of space and has repulsive gravity. The repulsive gravity is speeding up cosmic expansion. Dark energy accounts for about two-thirds of the mass-energy of everything, which means the major mass component of the universe was overlooked until 1998. For the discovery of dark energy, Saul Perlmutter, Brian Schmidt and Adam Riess won the 2011 Nobel Prize in Physics.

Nobody knows what dark energy is. In fact, when physicists use their best theory – quantum theory – to predict the energy of the vacuum, they get a number which is 1 followed by 120 zeroes bigger than observed. The biggest discrepancy between a prediction and observation in scientific history!

As the universe grows, it creates more dark energy with more repulsion, and the universe expands faster. Eventually, matter will be diluted out of existence. The universe may end not with a bang but with a whimper.

100% Cocoa
100% Cocoa
100% Cocoa

The third Big Bang bolt-on: inflation

Yet another way the basic Big Bang contradicts observations is harder to grasp. The Big Bang afterglow has essentially the same temperature everywhere in the sky. However, if we imagine cosmic expansion running backwards, to the point when the universe was 380,000 years old and the fireball radiation broke free of matter, parts of space which today are on opposite sides of the universe were not in contact. There had been insufficient time since the moment of creation for the fastest thing we know of – light – to travel between them. So, if one region cooled faster than another, heat could not flow in to equalize the temperature.

To explain the universe being at the same temperature, scientists postulate that, early on, it was far smaller than suspected. Then heat could travel everywhere. But, if the universe was smaller early on, it must have expanded faster to reach its current size in 13.82 billion years.

This super-fast expansion, known as 'inflation', occurred in the universe's first split-second. It was proposed independently by Alexei Starobinsky and Alan Guth around 1980.

Inflation has been likened to an H-bomb compared with the stick of dynamite of normal expansion that took over when 'inflation' ran out of steam. Inflation was driven by the vacuum. But no ordinary vacuum. A super-energetic vacuum with repulsive gravity.

During inflation, the universe expanded far faster than light. This is because space, in Einstein's theory of gravity, can expand at any rate it likes.

ATOMS
4.6%

DARK
MATTER
24%

DARK
ENERGY
71.4%

The ultimate free lunch

So here is the modern creation story . . . In the beginning was the inflationary vacuum. It had an extraordinary property. When it doubled its volume, it doubled its energy, when it tripled its volume it tripled its energy. Imagine if banknotes behaved like this. If you held a stack between your hands and pulled your hands apart, ever more banknotes would appear. No wonder physicists call inflation the 'ultimate free lunch'.

The inflationary vacuum expanded ever faster. But it was a 'quantum' thing. And quantum things like atoms and electrons are fundamentally unpredictable. So, randomly, all over the inflationary vacuum, bits 'decayed' into dull, low-energy, everyday vacuum. Think of tiny bubbles forming in a vast ocean. And in each bubble the enormous energy of the inflationary vacuum had to go somewhere. It went into creating matter and heating it to a blisteringly high temperature. It went into creating Big Bangs. Our Big Bang universe is one such bubble in the ever-expanding inflationary universe.

To start all this required an initial chunk of inflationary vacuum weighing only about a kilogram. Where did this come from? Incredibly, the laws of quantum theory permit energy to pop into existence out of nothing. The universe may therefore have been started by a piece of inflationary vacuum appearing out of . . . nothing.

The multiverse

Inflation suggests that the universe may effectively be infinite. However, we cannot see all of it for the simple reason that the universe has existed for only 13.82 billion years. This means that we can only observe those stars and galaxies whose light has taken less than 13.82 billion years to reach us. The light from objects whose light would take more than 13.82 billion years is still on its way.

Consequently, the 2 trillion galaxies we see are in a sphere of space known as the 'observable universe', which is bounded by a 'light horizon'. This horizon is like the horizon at sea. We know there is more ocean beyond the terrestrial horizon. And we know there is more of the universe beyond the cosmic horizon. Picture an infinite number of other regions of space like our observable universe.

What is it like in the other regions of this 'multiverse'? Well, according to inflation, the seeds of clusters of galaxies were random convulsions, or 'quantum fluctuations', of the early vacuum, which were magnified by the inflationary expansion. These would have led, in the other regions, to different galaxies and different stars – to *different histories*. In such an infinite universe everything that can happen happens. Out in the multiverse there may be copies of you living entirely different lives!

Where did the universe come from?

The twin pillars of modern physics are Einstein's theory of gravity – the general theory of relativity – and quantum theory. Einstein's theory describes stars and galaxies and the whole universe. Quantum theory is our best description of the world of atoms and their constituents. It has given us lasers and computers and nuclear reactors; it explains why the Sun shines and why the ground beneath our feet is solid.

Einstein's theory – which describes the large – and quantum theory – which describes the small – reign supreme in their own domains but have resisted any merger. This would not be a problem except that, once upon a time – in the Big Bang – the universe was very small. To understand how the universe emerged, like a rabbit from a hat, we therefore need to unite Einstein's theory with quantum theory: to create a quantum theory of gravity.

The best candidate for such a theory is 'string theory', which views the basic building blocks of reality as tiny strings of mass-energy vibrating like violin strings in ten-dimensional space-time. Nobody knows whether it will succeed. But one thing is certain. Only if we obtain a quantum theory of gravity will we be able to answer the ultimate questions: What is space? What is time? What is the universe? And where did it come from?

Further Reading

Stephen Hawking *A Brief History of Time* (Bantam, 1989)

Steven Weinberg *The First Three Minutes* (Basic Books, 1993)

Alan Guth *The Inflationary Universe* (Vintage, 1998)

Brian Greene *The Elegant Universe* (Vintage, 2000)

Marcus Chown *The Magic Furnace* (Vintage, 2000)

Simon Singh *Big Bang* (HarperPerennial, 2005)

Alex Vilenkin *Many Worlds in One* (Hill & Wang, 2007)

Marcus Chown *Afterglow of Creation* (Faber and Faber, 2010)